UNITED STATES
NAVAL POSTGRADUATE SCHOOL

OPERATIONAL ANALOGUES FOR KINETIC STUDIES

by

RICHARD A. REINHARDT
Assistant Professor of Chemistry

GILBERT F. KINNEY
Professor of Chemical Engineering

RESEARCH PAPER NO. 15
November 1957

OPERATIONAL ANALOGUES FOR KINETIC STUDIES

by

RICHARD A. REINHARDT

Assistant Professor of Chemistry

GILBERT F. KINNEY

Professor of Chemical Engineering

Research Paper No. 15

UNITED STATES NAVAL POSTGRADUATE SCHOOL

Monterey, California

November 1957

FOREWORD

This paper was presented, upon invitation of the symposium chairman, as part of the Beckmann Award Symposium at the 131st meeting of the American Chemical Society on 8 April 1957 in Miami, Florida.

ABSTRACT

Programming of the operational analogue computer for the rapid solution of problems dealing with complex chemical reaction-rate mechanisms is described. As an illustration, an enzyme-catalyzed reaction is considered and it is shown that computer results agree well with experiment, but not with the steady-state approximation customarily made for this system. It is pointed out that the operational analogue computer is a relatively simple and inexpensive set-up, and that programming principles, as developed and described here, make it applicable to a wide variety of theoretical and industrial problems.

An operational analogue computer is an assemblage of independent high-gain DC amplifiers, with accessory resistors, capacitors, and potentiometers, and a peg-board arrangement for making connections. Each time-dependent variable is represented on the computer as a voltage, and the amplifiers can either add together several such voltages, or else add and integrate. It is the purpose of this paper to show how an analogue computer of this type can be used for investigating the mechanism of a complex chemical reaction. The unravelling of such a mechanism would then, for example, permit ready scale-up from laboratory to plant size equipment, and thus be of valuable assistance in the selection of optimum size reactors.

In many actual applications of chemical kinetics, the rate law corresponding to a given mechanism is approximated by the use of simplifications suitable to the problem at hand - such as the steady-state hypothesis customarily required in a large number of chemical mechanism studies.[3] When, however, the concentration of each species (including intermediates) is desired as a function of time, especially in the early stages of reaction before the steady state is established, it becomes necessary to use the complete integrated form of the rate law. This form is also required if the rates of several steps are of comparable magnitude, so that the concept of a rate-determining step is not significant. These complete expressions are extraordinarily complex and, indeed, in many cases can be

evaluated only by numerical methods. This is especially true for reactions involving bimolecular or trimolecular steps in the mechanism.

An alternative treatment for the complete rate law is that of working with the set of simultaneous differential equations directly obtainable from the mechanism.[6] Each of these equations is of first order and first degree, though non-linear for bimolecular or trimolecular steps. (It should be noted that even though the reaction is of higher than first order in the chemical sense, each differential equation is mathematically of the first order.)

Consider, for example, a reaction of two consecutive unimolecular steps:

$$A \xrightarrow{k_1} B \xrightarrow{k_2} C \qquad (1)$$

Selecting the rate equation for component B as an example, and using lower-case letters to represent concentration,

$$\frac{db}{dt} = k_1 a - k_2 b \qquad (2)$$

Defining the operator 1/P by the equation

$$1/P \left[F(t) \right] = \int_0^t F(t)dt, \qquad (3)$$

equation 2 may be integrated to

$$b = \int_0^t (k_1 a - k_2 b)dt + b_0 \qquad (4)$$

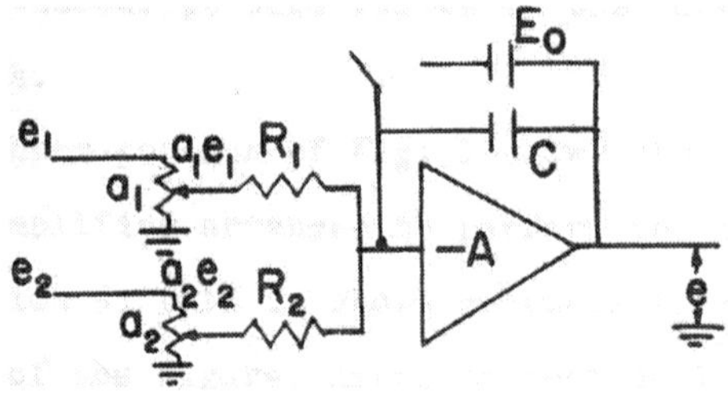

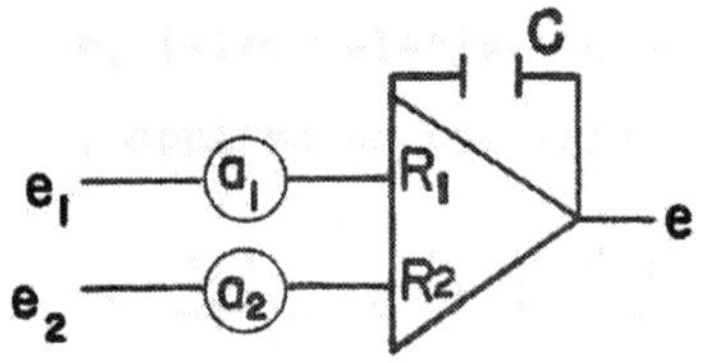

FIG. 1 THE INTERGRATING AMPLIFIER: (upper) ELECTRICAL CIRCUIT; (lower) SCHEMATIC.

or $$b = -1/P\,(k_2 b - k_1 a) + b_0 \qquad (5)$$

in which subscript zero refers to the concentration at zero time.

The upper portion of Fig. 1 shows the electrical circuit for an amplifier arranged to perform the operation indicated by equation 5; this is shown schematically in the lower portion of the figure, using conventional computer symbols.[7] The amplifier, with amplification -A, is shown as a triangle, the grid input being to the base of the triangle, the plate output from the apex. The potentiometers, set at a_1 and a_2, control the fraction of the input voltages, e_1 and e_2 (referred to ground) fed through the input resistors, R_1 and R_2, to the grid of the amplifier. For feed-back capacitance C and with output voltage e, (also relative to ground) it follows from Kirchhoff's Law, applied to the grid input, that

$$\frac{a_1 e_1 - e_g}{R_1} + \frac{a_2 e_2 - e_g}{R_2} + C\,\frac{d(e - e_g)}{dt} + i_g = 0, \qquad (6)$$

where subscript g refers to the grid. Now the grid voltage and current are very small for a well-designed high-gain amplifier, and from this consideration it develops[5,7,] that, in terms of the operator 1/P, the output voltage becomes

$$e = -1/P\left(\frac{a_1}{R_1 C}\,e_1 + \frac{a_2}{R_2 C}\,e_2\right) + E_0, \qquad (7)$$

where E_0 is the initial value of the output voltage (disconnected when computer operation begins).

or [illegible] $= -1/P$ [illegible] (5)

in which [illegible] zero time.

The upper portion [illegible] for an amplifier [illegible] by symbols [illegible] portion of the figure, [illegible] The amplifier, which [illegible] the grid input [illegible] output from [illegible] control [illegible] to ground) fed [illegible] grid of the amplifier. [illegible] output voltage [illegible] Kirchhoff's Law, [illegible] grid [illegible]

$$[illegible] \qquad (6)$$

where [illegible] the grid. Now [illegible] and current are very [illegible] amplification [illegible] in terms of the [illegible] the [illegible] voltage [illegible]

$$[illegible] \qquad (7)$$

where [illegible] of the output voltage (disconnected when [illegible]).

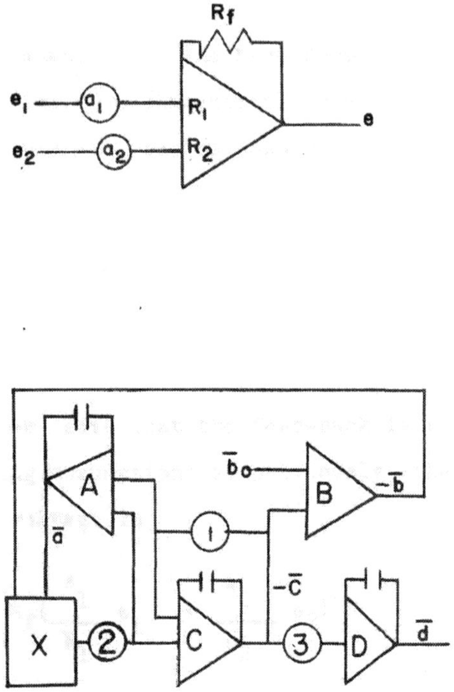

FIG. 2 THE SUMMING AMPLIFIER (upper)
SET-UP FOR THE ENZYME PROBLEM. (lower)

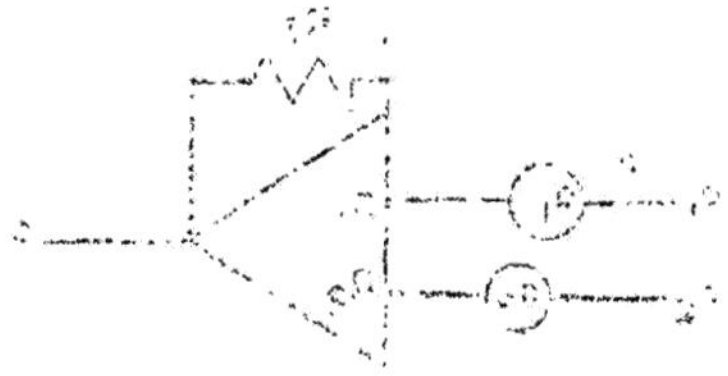

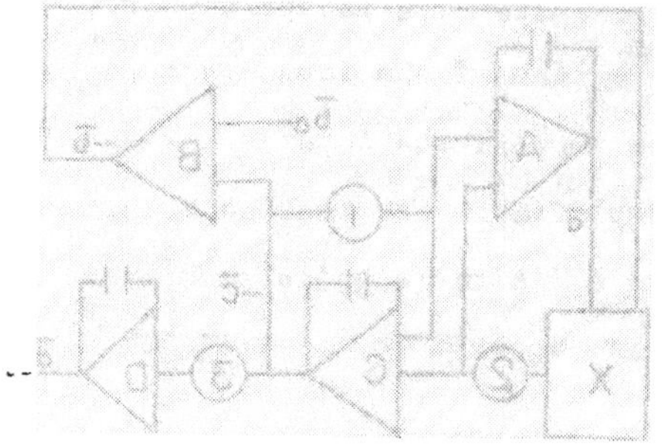

FIG.2 THE SUMMING [illegible]

SET-UP FOR THE [illegible]

The similarity in form between equations 5 and 7 suggests the use of such amplifiers in solving the type of problem exemplified by equation 1. The requirement is that the voltages suitably represent concentrations and the resistor-capacitor-potentiometer combinations represent the rate constants.

In many problems it is also necessary to add together variables. For this purpose can be used the summing amplifier, as shown schematically in the upper portion of Figure 2. The arrangement is similar to that for the integrating amplifier, save that the feed-back is resistive, of value R_f. Making assumptions as previously, the equation for the output voltage is

$$e = -R_f\left(\frac{a_1}{R_1} e_1 + \frac{a_2}{R_2} e_2\right) \tag{8}$$

For single input and equal resistors, the summing amplifier functions as a sign changer.

If the mechanism under consideration contains steps of molecularity greater than one, it is also necessary to use a function multiplier, which is an electronic device to generate the product of two variable voltages. For commercially available multipliers the output, as a function of input voltages e_1 and e_2, is given by

$$e = -me_1e_2, \tag{9}$$

The similar [illegible]

the use of such [illegible]

exemplified by [illegible]

tages suitably [illegible]

capacitor-pote[illegible]

capacitance.

In many prob[illegible]

variables. [illegible]

plifier, as sh[illegible]

Figure 2. The [illegible]

erating amplifi[illegible]

value R_f. [illegible]

for the output [illegible]

[illegible]

For single input [illegible]

functions as a [illegible]

If the mech[illegible]

molecularity [illegible]

a function mult[illegible]

generate the pr[illegible]

cially availab[illegible]

input voltages [illegible]

where m is a constant of proportionality (equal to 0.01 for the equipment used in this laboratory). On the schematic diagrams the multiplier is shown as a rectangle with an X at its center. Another auxiliary piece of equipment is a function generator, whose output conforms to some arbitrarily given value for any given input. This device is useful for non-isothermal reactions, where rate constants may change.

To illustrate the use of computer technique, the following mechanism has been selected:

$$A + B \underset{k_2}{\overset{k_1}{\rightleftharpoons}} C \xrightarrow{k_3} B + D \qquad (10)$$

in which B is a catalyst of some sort. Equation 10 could, for example, represent the Michaelis-Menten formulation for enzyme activity[4], B symbolizing the enzyme and A the substrate. Assuming only A and B present initially, the operator equations for A, C, and D become

$$a = -1/P\left[k_1 ab + k_2(-c)\right] + a_0 \qquad (11)$$

$$-c = -1/P\left[k_1 ab + k_2(-c) + k_3(-c)\right] \qquad (12)$$

$$d = -1/P\left[k_3(-c)\right] \qquad (13)$$

The equations are written in terms of -c to avoid the use of additional amplifiers for sign changing. The equation diagrams the

at its center.

function gener

given value

non-isotherm

To illustrat

mechanism bas

A + B

in which B is

for example,

enzyme activi

strate. Assu

operator equa

a =

-c =

d =

The equations

of additional

for B can be deduced from the stoichiometry of the system:

$$-b = -\left[b_0 + (-c) \right] \tag{14}$$

-b being used here since the function multiplier changes sign. These four equations suggest the use of at least four amplifiers (three integrating and one summing) plus a function multiplier. The arrangement used to solve this problem is shown schematically in the lower portion of Figure 2.

The specific problem chosen for illustration was the enzyme-catalyzed oxidation of an appropriate oxygen acceptor by hydrogen peroxide, a system which has been studied by Chance[1]. Typical values of the parameters for this problem are:

$a_0 = 4 \times 10^{-6}$ $\underline{M}$, $b_0 = 1 \times 10^{-6}$ $\underline{M}$, $k_1 = 1 \times 10^7$, $k_2 = 0.2$, $K_3 = 5$

(t in seconds for the rate constants). To illustrate the scaling, it will suffice to consider only the amplifier generating $\bar{a}$, the voltage representing a. Let $\lambda_A = a/\bar{a}$, $\lambda_B - b/\bar{b}$ $\lambda_C = c/\bar{c}$, where each λ is a magnitude-scaling factor, and $\bar{a}$, $\bar{b}$, and $\bar{c}$ the voltages representing the respective concentrations. Then equation 11 may be rewritten

$$\bar{a} = -1/P \left[k_1 \lambda_B \overline{ab} + k_2 \frac{\lambda_C}{\lambda_A} (-\bar{c}) \right] + a_0 \tag{15}$$

To take care of time-scaling, a factor β is introduced with each feed-back capacitance, such that β seconds on the computer corresponds to one second of actual time. Then by

applying the electrical equations 7 and 9 to this amplifier,

$$\bar{a} = -1/P\left[\frac{a_1 m\beta}{R_1C}\,\overline{ab} + \frac{a_2\beta}{R_2C}(-\bar{c})\right] + a_0 \tag{16}$$

where subscript 1 refers to the portion of the circuit into which $m\overline{ab}$ is fed, subscript 2 to that for $-\bar{c}$. By equating coefficients in equations 15 and 16,

$$k_1 \lambda_B = \frac{a_1 m\beta}{R_1C} \tag{17}$$

$$k_2 \frac{\lambda_C}{\lambda_A} = \frac{a_2\beta}{R_2C} \tag{18}$$

These two equations, combined with similar equations for the other amplifiers, permit the proper selection of the resistors, capacitors, and potentiometer settings. In Table 1 are shown values of these components corresponding to one case considered, and in Figure 3 are shown simultaneous traces (from a four-channel recorder) of each of the concentrations _vs_ time for this case. Concordance with the results of Chance[1] are quite satisfactory: _e.g._, the maximum concentration of the intermediate C and the time required to attain this maximum agree quite well with the results reported by him.

The assumption that there is established a steady state in b and c, after an initial transient period[4], leads to the relationship

Table I. Typical Values of Components for the Enzyme Problem

	Amplifier (Species)			
	A	B	C	D
Magnitude-scaling factor, λ, x 10^8	10	1	1	10
Feed-back	1μ	.2 meg	1μ	1μ
Input Resistors, megohms	1, 10	.2	.1, 1, 1	1
Initial voltage	38	95	0	0
Initial concentration, micromoles/liter	3.8	.95	0	0

	Potentiometer (Rate-constant)		
	1	2	3
Potentiometer Setting	1.0	0.02	0.5
Rate Constant (t in sec.)	1×10^7	0.2	5

Time-scaling factor, β, = 10

Multiplier scaling factor, m, = 0.01

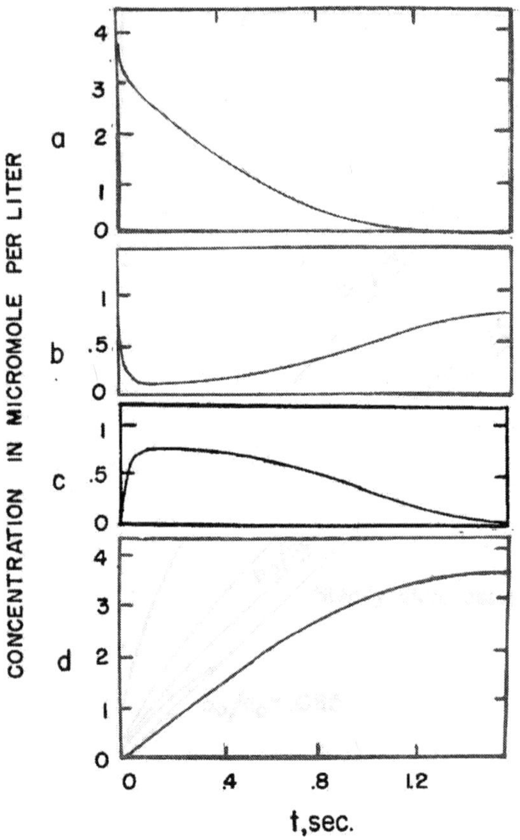

FIG.3 KINETICS OF THE ENZYME PROBLEM.

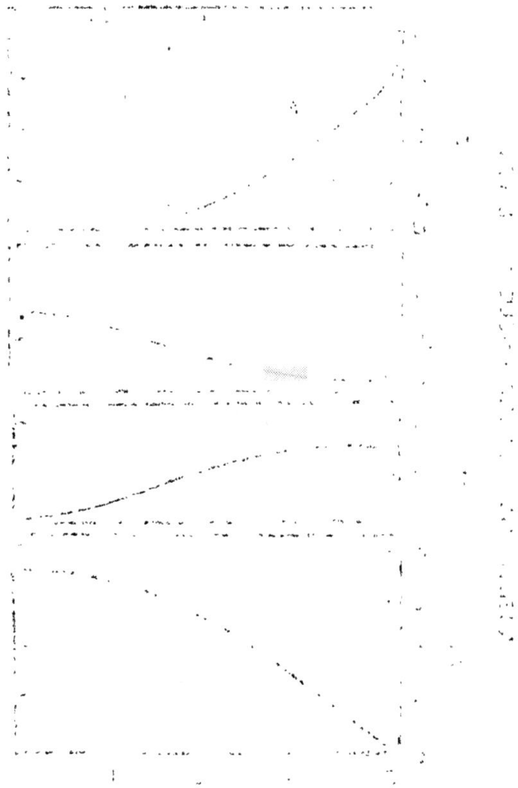

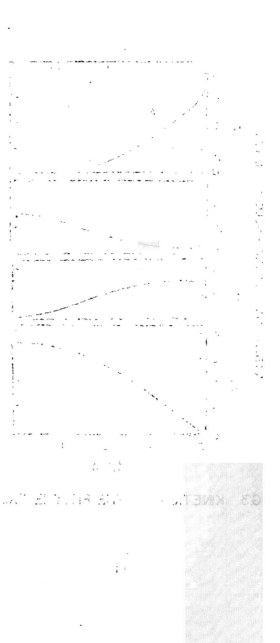

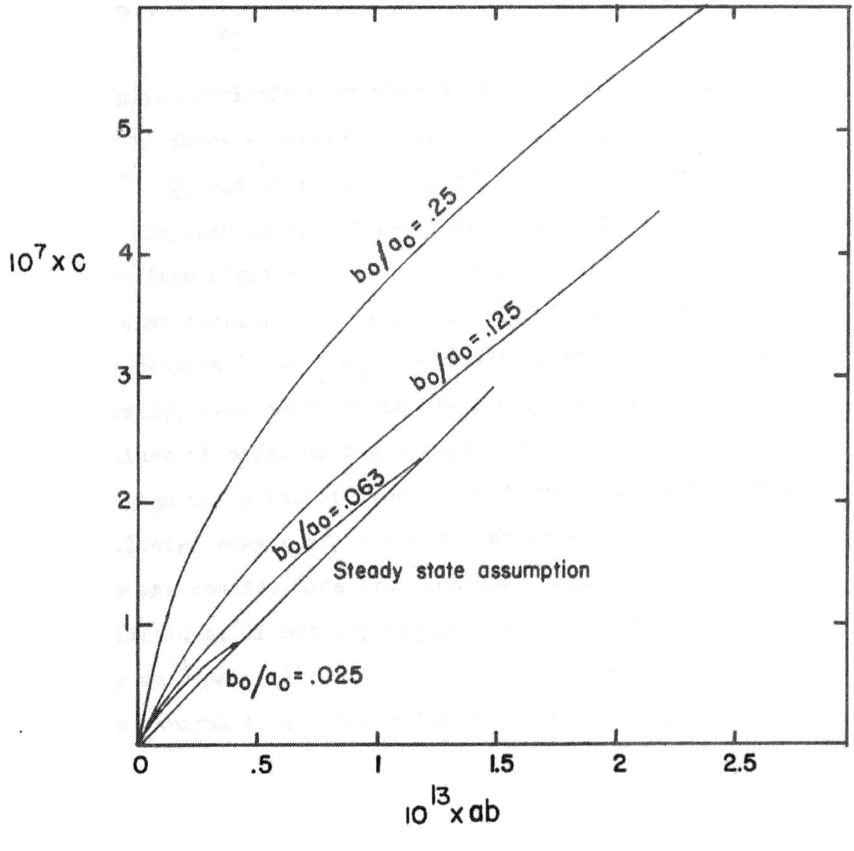

FIG. 4 EFFECT OF CATALYST CONCENTRATION ON STEADY—STATE ASSUMPTION.

$$c = \frac{k_2 + k_3}{k_1} ab, \qquad (19)$$

which applies strictly only when db/dt = dc/dt = 0. In Figure 4 is shown a portion of the plot of c vs. ab, for $a_0 = 4 \times 10^{-6}$ M, and various values of b_0, as made directly from the computer on an X-Y recorder. (note that time increases from right to left in this figure.) The straight line is that computed from equation 19. It can be seen that if b_0 is comparable with a_0, the steady-state approximation is not valid, even near the completion of reaction. Only at small values of b_0/a_0 do the actual curves approach equation 19, although the straight line could serve as a rough average for the latter stages of reaction. Whereas deductions such as these are readily made from computer traces, it is indeed quite difficult, if not impossible, to produce them by ordinary mathematical methods.

For a general kinetics problem at least one amplifier will be required for each component participating in the mechanism, plus such others as may be required for sign-changing; one potentiometer for each rate-constant; and one function multiplier for each bimolecular step (or two for a trimolecular step, etc.). The speed, simplicity, and adaptability of the operational analogue computer makes it also a useful device for the determination, from observed

kinetics data, of the rate constants for assumed mechanisms. Trial values of the various rate constants are used, by changing potentiometer settings and, if necessary, resistors, until a fit with the data is obtained. Since changing parameters ordinarily takes only a few seconds, such a problem can be solved quite rapidly.

It follows that the independent variable of the problem need not be time, but could also be any quantity varying linearly with time, using appropriate scaling. It is thus possible to study the kinetics of steady-flow processes. An example is the study of catalytic isomerization[2]: the results of mechanisms assuming either single-site or dual site adsorption are readily obtained for a wide variety of initial conditions and flow rates.

A computer of the sort described is of only moderate expense - a quite satisfactory combination need cost no more than a good polarimeter - and can be operated by a single chemist or chemical engineer working in his own laboratory. A problem of the type described can be set up and solved in a few hours.

LITERATURE CITED

(1) Chance, B., J. Biol. Chem. 151, 553 (1943)

(2) Corrigan, T.E., Chem. Eng. 62, No. 2, 195 (1955)

(3) Frost, A.A., and Pearson, R.G., "Kinetics & Mechanism", John Wiley, New York, 1955.

(4) Shaw, W.H.R., J. Chem. Ed. 34, 22 (1957)

(5) Wass, C.A.A., "Introduction to Electronic Analogue Computers", McGraw-Hill, New York, 1955.

(6) Wheeler, R.C.H., and Kinney, G.F., IRE Trans. on Industrial Electronics, p. 70 (March 1956)

(7) Korn, G.H., and Korn, T.M., "Electronic Analog Computers", 2d ed., McGraw-Hill, New York, 1956.

(8) See also Rose, A., et al., Ind. Eng. Chem. 48, 627 (1956)

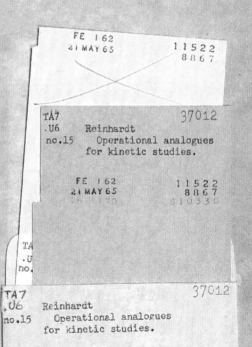

www.ingramcontent.com/pod-product-compliance
Lightning Source LLC
LaVergne TN
LVHW012201040326
832903LV00003B/46

* 9 7 8 1 0 1 9 3 6 6 7 0 7 *